GLASS: A POCKET DICTIONARY OF TERMS COMMONLY USED TO DESCRIBE GLASS AND GLASSMAKING

Compiled By
David Whitehouse

The Corning Museum of Glass
Corning, New York

Cover: Trick glass. Spain, Barcelona, 17th century. H. 21.9 cm.
Gift of The Ruth Bryan Strauss Memorial Foundation.

Copyright © 1993
The Corning Museum of Glass
Corning, New York 14830-2253

Design and Typography: Graphic Solutions
Printing: Cayuga Press

Standard Book Number 0-87290-132-7
Library of Congress Catalog Card Number 93-072859

FOREWORD

This short dictionary is intended to help students and collectors of glass to understand some of the unfamiliar words that they may encounter in books, catalogs, and museum labels. It contains definitions of words and phrases that describe glassmakers' materials, techniques, tools, and products. Words that appear in **boldface** type in the definitions are defined elsewhere in the dictionary.

The illustrations have been selected from objects in The Corning Museum of Glass, and from books and other materials in the Museum's Rakow Library. They are described on pages 86 and 87.

Many of the definitions in the dictionary are based on the glossary in the catalog that accompanied the exhibition "Treasures from The Corning Museum of Glass," which was shown at the Yokohama Museum of Art in 1992. The catalog was produced jointly by the Corning and Yokohama museums, and we are grateful to our Japanese colleagues for allowing the glossary to be revised and enlarged for this publication.

I am indebted to Robert H. Brill, Susanne K. Frantz, Dwight P. Lanmon, and Jane Shadel Spillman for numerous improvements to the text. Needless to say, I am solely responsible for any errors that remain.

David Whitehouse
Director
The Corning Museum of Glass

Acid-etched vase. France, Maurice Marinot, 1934. H. 17.1 cm.

A

Abrasion: The technique of grinding shallow decoration with a wheel. The decorated areas are left unpolished.

Acanthus: (1) A group of Mediterranean, Asian, and African plants with large, spiny leaves; hence, (2) ornament that resembles the leaves of the species *Acanthus spinosus.*

Acid etching: The process of etching the surface of glass with **hydrofluoric acid**. Acid-etched decoration is produced by covering the glass with an acid- resistant substance such as wax, through which the design is scratched. A mixture of dilute hydrofluoric acid and potassium fluoride is then applied to etch the exposed areas of glass. An effect superficially similar to **weathering** may be obtained by exposing glass to fumes of hydrofluoric acid to make an allover matte surface.

Acid polishing: The process of making a glossy, polished surface by dipping the object, usually of cut glass, into a mixture of **hydrofluoric** and sulfuric acids.

Acid stamping: The process of **acid etching** a trademark or signature into glass after it has been **annealed,** using a device that resembles a rubber stamp.

Aeolipile (Greek): The name sometimes given to globular or pear-shaped objects with a narrow neck and mouth. The function of these objects is uncertain. The word was originally applied to a device, invented in the second century B.C., in which a closed, water-filled vessel, when heated, was made to rotate by jets of steam issuing from one or more projecting, bent tubes. Most surviving aeolipiles, however, are Islamic; they are believed to be containers. See also **Grenade.**

Agate glass: See *Calcedonio.*

Air trap, air lock: An air-filled void, which may be of almost any shape. Air traps in stems are frequently tear-shaped or spirally twisted. See **Diamond air trap** and **Twist.**

Air twist: See **Twist.**

Alabaster glass: A type of translucent white glass, similar to **opal glass**, first produced in Bohemia in the 19th century. In the 1920s, Frederick Carder (1863-1963) introduced alabaster glass at Steuben Glass Works in Corning, New York. Carder's alabaster glass has an iridescent finish made by spraying the object with stannous chloride and then reheating it.

Alabastron (Greek), **alabastrum** (Latin): A small bottle or flask for perfume or toilet oil, usually with a flattened rim, a narrow neck, a cylindrical body, and two small handles.

Ale glass: A type of English drinking glass for ale or beer. Ale glasses, first made in the 17th century, have a tall and conical cup, a stem, and a foot. They may be enameled, engraved, or gilded with representations of hops or barley.

Alembic (Arabic *al-anbiq*, "the still"): An apparatus used for distilling.

Alkali: A soluble salt consisting mainly of potassium carbonate or sodium carbonate. It is one of the essential ingredients of **glass**, generally accounting for about 15-20 percent of the **batch**. The alkali is a **flux**, which reduces the melting point of the major constituent of glass, **silica**.

Almorrata (Spanish): A rose water sprinkler with many spouts, made in northern Spain between the 16th and 18th centuries.

Amberina: A type of **Art Glass** that varies in color from

Amphoriskos. Eastern Mediterranean, 2nd-1st century B.C. H. 24 cm.

amber to ruby red or purple on the same object. This shaded effect is due to the presence of gold in the **batch**. The object is amber when it emerges from the **lehr**, but partial reheating causes the affected portion to become red or purple. Amberina, developed by Joseph Locke (1846-1936) at the New England Glass Company in East Cambridge, Massachusetts, was patented in 1883.

Amen glass: A rare type of English wineglass with a drawn stem. The bowl is decorated by **diamond-point engraving** with verses from the Jacobite Hymn followed by the word "Amen," and with emblems associated with the Jacobite uprising of 1715. See **Jacobite glass**.

Amphora (Latin): A jar with two handles.

Amphoriskos (Greek, "small amphora"): A small jar with two handles, used for perfume or toilet oil in the pre-Roman and Roman periods.

Amulet: A charm believed to protect the wearer against evil or to bring good fortune.

Ancient glass: A term frequently used to mean all pre-Roman and ancient Roman glass.

Annagrün (German): A type of yellowish green glass colored by adding uranium oxide to the **batch**. Developed by Josef Riedel (1816-1894), who named it for his wife, Anna, this glass was made from the 1830s and 1840s. See also **Uranium glass**.

Annealing: The process of slowly cooling a completed object in an auxiliary part of the glass **furnace**, or in a separate furnace. This is an integral part of glassmaking because if a hot glass object is allowed to cool too quickly, it will be highly strained by the time it reaches room temperature; indeed, it may break as it cools. Highly strained glasses break easily if subjected to mechanical or thermal shock. See **Lehr**.

Applied decoration: Heated glass elements (such as **canes**, *murrini*, and **trails**) applied during manufacture to a glass object that is still hot, and either left in relief or **marvered** until they are flush with the surface. See also **Marquetry** and **Pick-up decoration**.

Arabesque: (1) In Islamic art, an intricate pattern of interlaced ornament consisting of curvilinear stems and tendrils that terminate in leaves; (2) in Renaissance and later European art, a pattern of interlaced curvilinear stems, scrolls, and leaves, sometimes containing animal motifs.

Argand lamp: An oil-burning lamp with a glass **chimney**, named for the Swiss physicist and inventor Aimé Argand (1750-1803), who invented the tubular wick burner in 1782. Argand lamps are efficient because the tubular wick feeds oxygen to the flame and the chimney increases the draft.

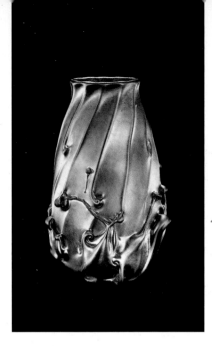

Art Nouveau vase. U.S., Steuben, about 1910. H. 17.1 cm.

Art Glass: (1) Several types of glass with newly developed surface textures, shaded colors, or **casing**, made in the United States from about 1870 and in Europe between about 1880 and 1900; (2) more generally, any ornamental glassware made since the mid-19th century.

Art Nouveau (French, "new art"): An international, late 19th- and early 20th- century decorative style characterized by organic foliate forms, sinuous lines, and non-geometric, "whiplash" curves. Art Nouveau originated in Europe in the 1880s, and reached the peak of its popularity around 1900. In America, it inspired, among others, Louis Comfort Tiffany (1848-1933). The name is derived from *"La Maison de l'Art Nouveau,"* a gallery for interior design that opened in Paris in 1896. The German term for Art Nouveau is ***Jugendstil***.

Aryballos (Greek): A small globular flask with two handles, used by the ancient Greeks and Romans to contain toilet oil.

Baluster. England, about 1695. H. 24.7 cm.

At-the-fire: The process of reheating a blown glass object at the **glory hole** during manufacture, to permit further inflation and/or manipulation with tools.

At-the-flame (at-the-lamp, lampworking): See **Flameworking**.

Aurene glass: A type of ornamental glass with an **iridescent** surface made by spraying the glass with stannous chloride or lead chloride and reheating it under controlled atmospheric conditions. Aurene glass was developed by Frederick Carder (1863-1963) at Steuben Glass Works in Corning, New York, in 1904.

Aventurine (from French *aventure*, "chance"): Translucent glass with sparkling inclusions of gold, copper, or chromic oxide, first made in Venice in the 15th century. Aventurine glass imitates the mineral of the same name, a variety of quartz spangled with mica.

B

Baluster: A type of English drinking glass of the late 17th and 18th centuries, with the stem in the form of a baluster. (In architecture, a baluster is a short vertical support with a circular section and a vaselike outline.)

Balustroid: A variety of **baluster** glass with an elongated stem, current in England between about 1725 and 1760.

Bandwurmglas (German, "tapeworm glass"): A variety of ***Stangenglas*** decorated with a notched **trail** wound spirally, like a worm, around the wall. Glasses of this type were made in Germany between the 15th and 17th centuries.

Bar: A single piece of glass formed by fusing several **canes** or **rods**. A bar can be cut into numerous slices, all with the same design, to be used as **inlays** or appliqués, or in making **mosaic glass**.

Barilla: (1) A plant, *Salsola soda*, which grows extensively on seashores in Spain, Sicily, and the Canary Islands; hence (2) an impure **alkali** made by burning plants of this and related species, formerly used in the manufacture of soda, soap, and **glass**.

Batch: The mixture of raw materials (often **silica**, **soda**, **potash**, **lime**, and **cullet**) that is heated in a **pot** or **tank** to make **glass**.

Battledore: A glassworker's tool in the form of a square wooden paddle with a handle. Battledores are used to smooth the bottoms of vessels and other objects.

Bear jar: A 19th-century American pressed glass jar in the form of a bear, probably for bear grease.

Berkemeyer (German): A type of drinking glass, similar to a ***Römer***, but with a funnel-shaped mouth. It was made in

Germany and the Low Countries in the 16th and 17th centuries.

Berry set: A large bowl with matching smaller bowls, used for serving fruit and other desserts.

Biedermeier beakers. Austria, Vienna, about 1811-1828. H. (tallest) 11 cm.

Biedermeier style: A style of decorative art favored by the German middle class between about 1820 and 1840. The name is derived from two fictional bourgeois characters, Biedermann and Bummelmeier, in the satirical verses of Ludwig Eichrodt. During the period in which the Biedermeier style was popular, glassmaking revived in Bohemia, where new kinds of glass such as **Lithyalin** and elaborate **flashed**, **wheel-engraved**, and **enameled** glass were produced for middle-class consumers.

Bird fountain: A **flameworked** centerpiece or mantel ornament consisting of a tall fountain with two birds perched on the rim, and two or more shorter pedestals, each with a bird on the top. The birds have tails made of glass fibers. Bird fountains were made in England in the mid-19th century.

Bitters bottle: A bottle for bitters, alcoholic beverages flavored with bitter herbs. Bitters, sold as medicine rather than as liquor (and for this reason, taxed more leniently), were immensely popular in America in the second half of the 19th century.

Black bottle: A popular term for bottles of dark green or dark brown glass, the dark color of which protected the contents. "Black" glass was first made in England in the mid-17th century.

Blank: Any cooled glass object that requires further forming or decoration to be **finished**.

Blankschnitt (German, "polished cut"): A style of engraved decoration in which the relief effect is enhanced by **polishing** the ground part of the **intaglio**. *Blankschnitt* decoration is frequently found on glasses engraved in the German city of Nuremberg in the 17th and 18th centuries.

Bleeding glass: See **Cupping glass**.

Blobbing: The technique of decorating hot glass by dropping onto the surface blobs of molten glass, usually of a different color or colors.

Block: A block of wood hollowed out to form a hemispherical recess. After it has been dipped in water to reduce charring and to create a "cushion" of steam, the block is used to form the **gather** into a sphere, prior to inflation.

Blowing: The technique of forming an object by inflating a gob of molten glass **gathered** on the end of a **blowpipe**. The **gaffer** blows through the tube, slightly inflating the gob, which is then manipulated into the required form by swinging it, rolling it on a **marver**, or shaping it with tools or in a **mold**; it is then inflated to the desired size.

Blown three-mold glass: Glassware made in America between about 1815 and 1835 that was blown in a full-size mold that (despite the popular name) consisted of between two and five pieces.

Blowpipe: An iron or steel tube, usually about five feet long, for blowing glass. Blowpipes have a mouthpiece at one end and are usually fitted at the other end with a metal ring that helps to retain the **gather**.

Borsella (Italian): A tonglike tool used for shaping glass. The *borsella puntata* has a pattern on the jaws, which is impressed on the glass.

Bottle glass: A common, naturally colored, dark greenish or brownish glass. The color is characteristic of glass that includes traces of iron found in the natural **silica** used as the major ingredient. Sometimes, additional iron, in the form of iron oxide (or other materials), is added to darken the color.

Bow lathe: A primitive lathe powered by the use of a bow. The bowstring is looped around the spindle of the lathe and causes it to rotate as the bow is drawn backward and forward.

Breakfast set: A sugar bowl and matching **creamer**.

Brilliant-cut glass: Objects with elaborate, deeply cut patterns that usually cover the entire surface and are highly polished. In the United States, the vogue for brilliant-cut glass lasted from about 1880 to 1915.

Broad glass: See **Cylinder glass**.

*Burmese lamp. U.S.,
about 1885-1895.
H. 48.5 cm.*

Broken-swirl ribbing: Mold-blown decoration that has
two sets of ribs. This is made by blowing the **gather** in a
vertically ribbed **dip mold**, extracting and twisting it to
produce a swirled effect, and then redipping it in the same
or another dip mold to create a second set of ribs.

Bubble: A pocket of gas trapped in glass during
manufacture. The term is used for both bubbles
introduced intentionally (also known as **air traps** or beads)
and bubbles trapped accidentally during the melting
process. Very small bubbles are known as **seeds**.

Bull's-eye pane: A glass **pane** with a pontil mark
surrounded by concentric ridges. This was the central
part of a large pane of **crown glass**.

Burmese glass: A type of translucent pink-shading-to-
yellow **Art Glass** made by the Mount Washington Glass
Company in New Bedford, Massachusetts, between 1885
and about 1895.

Burner: The part of a lamp where the flame is produced.

Burning fluid: A mixture of alcohol and turpentine, used as lamp fuel in the 19th century. Burning fluid, which was dangerously explosive, was replaced by kerosene in the late 1850s.

C

Cable: A pattern resembling the twisted strands of a rope.

Caddy: A small, lidded container, usually for tea.

Cage cup: An ancient Roman vessel decorated by **undercutting** so that the surface decoration stands free of the body of the glass, supported by struts. The vessel appears, therefore, to be enclosed in an **openwork** cage. Cage cups are sometimes known as **diatreta** or **vasa diatreta**.

Calcedonio (Italian, "chalcedony"): Glass marbled with brown, blue, green, and yellow swirls in imitation of chalcedony and other banded semiprecious stones. *Calcedonio* was first made in Venice in the late 15th century.

Came: A grooved strip of lead or (rarely) another metal, generally with an H- shaped cross section, used to join separate parts of glass windows.

Cameo glass: Glass of one color covered, usually by **casing,** with one or more layers of contrasting color(s). The outer layers are **acid-etched, carved, cut,** or **engraved** to produce a design that stands out from the background. The first cameo glasses were made by the ancient Romans. The genre was revived in England and, to a lesser extent, in America in the late 19th century.

Cameo glass plaque. England, George Woodall, 1898.
D. 46.3 cm.

Candelabrum: A candle holder or lamp with several arms or branches.

Candlestick: An object with a socket or spike for one candle.

Cane: A thin, monochrome rod, or a composite rod consisting of groups of rods of different colors, which are bundled together and fused to form a polychrome design that is visible when seen in cross section. See **Bar**, **Millefiori**, and **Rod**.

Cántaro (Spanish): A drinking vessel shaped like a closed pitcher, with a ring handle at the center and two spouts, a short one for filling and pouring, and a longer one through which the beverage can be poured into the drinker's mouth.

Carnival glass: Inexpensive pressed glass with vivid gold, orange, and purple iridescence, made in the United States between about 1895 and 1924. It is so called because it was frequently offered as fairground prizes.

Carving: The removal of glass from the surface of an object by means of hand-held tools.

Casing: The application of a layer of glass over a layer of contrasting color. The **gaffer** either gathers one layer over another **gather**, or inflates a gob of hot glass inside a preformed **blank** of another color. The two components adhere and are inflated together (perhaps with frequent reheating) until they have the desired form. Sometimes, the upper layer is **carved**, **cut**, or **acid-etched** to produce **cameo glass**.

Casting: The generic name for a wide variety of techniques used to form glass in a **mold**.

Castor: A small vessel with a perforated top from which one casts or sprinkles sugar or condiments such as pepper. A castor set is a matching group of castors, which, together with **cruets**, form a **condiment set**.

Celery handle: A handle with vertical ribbing like a celery stalk.

Celery vase: A tall, narrow vase used at the table for holding celery.

Chair: (1) The bench used by the **gaffer** while forming a glass object. Traditionally, this is a wide bench with arms, on which the gaffer rests the **blowpipe** with its **parison** of molten glass and rolls it backward and forward so that the parison retains its symmetrical shape during the forming process. (2) The team of glassworkers who assist a gaffer.

Chalk glass: A colorless glass containing chalk, developed in Bohemia in the late 17th century. Vessels of thick chalk glass were often elaborately engraved.

Chandelier (French, "candlestick"): A lighting fixture suspended from the ceiling, with two or more arms

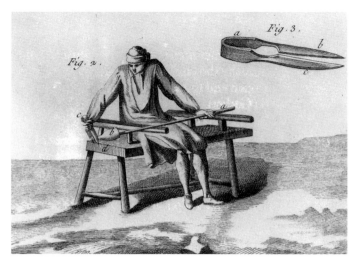

Chair. From Diderot, Receuil de planches..., *about 1775.*

bearing lights (originally, candles) or two or more pendent lights. Many chandeliers have faceted **lead glass** arms, candle cups, shafts, and prisms, which reflect the light and sparkle like tiny mirrors.

Chimney: A glass tube, open at both ends, used to shield the flame of an oil lamp, to trap soot, and to increase the draft.

Cinerary urn: A vessel for cremated human remains. In the Roman period, cremation was a widely used method of disposing of the dead, whose ashes were sometimes placed in glass cinerary urns.

Cintra glass: A type of decorative glass developed by Frederick Carder (1863-1963) at Steuben Glass Works in Corning, New York, before 1917. Most Cintra glass was made by picking up chips of colored glass on the **parison** and then **casing** them with a thin layer of (usually) colorless glass.

Cire perdue (French, "lost wax"): See **Lost wax casting**.

Clamp: A tool sometimes used instead of a **pontil** to hold the closed end (usually the bottom) of a partly formed glass vessel while the open end (usually the mouth) is being shaped. See also **Gadget**.

Clapper: A tool consisting of two rectangular pieces of wood joined at one end by a leather hinge. There is an aperture in one of the pieces of wood, and this holds the stem of a **goblet** or wineglass while it is being made. The clapper is used to squeeze a blob of glass in order to form the foot.

Claw beaker: A beaker decorated with claw- or trunklike protrusions made by applying blobs of hot glass that melted the parts of the wall to which they were attached. The blobs were then blown outward and manipulated to resemble hollow claws. Claw beakers were made in Europe between the fifth and seventh centuries A.D. Similar decoration was made in Germany in the 16th century.

Clichy rose: A slice of a **cane** depicting an open rose. Canes of this type were frequently used in **paperweights** made at the Clichy factory in France in the 19th century.

Clutha glass: A type of glass with **air traps** and specks of **aventurine**, patented in the 1890s by James Couper, Christopher Dresser, and George Walton.

Cluthra glass: A type of glass developed in the 1920s by Frederick Carder (1863-1963) at Steuben Glass Works in Corning, New York.

Coil base: A **trail** or thread of glass drawn out to form a ring or conical foot on which the vessel stands.

Cold painting. U.S., Robert Carlson, 1986. H. 64.5 cm.

Coin weight: The term popularly applied to Islamic coin-shaped weights or tokens, most of which were made in Egypt between the eighth and 12th centuries.

Cold colors: Pigments applied as decoration to glass by **cold painting**.

Cold painting: The technique of decorating an object by applying paint such as artists use on other materials. This is in contrast to enameling, in which powdered glasses of various colors are fused to the surface by heating. See also **Enamel**.

Cold working: The collective term for the many techniques (such as **copper-wheel engraving**) used to alter or decorate glass when it is cold.

Collar: (1) A band of applied glass around the rim of a vessel. On bottles, the collar is used to secure the cork. (2) A threaded metal ring around the **font** of a lamp, used to attach a screw-in burner.

Colored glass: Glass that is colored by (1) impurities in the basic ingredients in the **batch** or (2) techniques of coloring glass by one of three main processes: (a) using a dissolved **metallic oxide** to impart a color throughout, (b) forming a dispersion of some substance in a colloidal state, and (c) suspending particles of pigments to form opaque colors.

Combed decoration. U.S., about 1860-1870. H. 15.3 cm.

Combed decoration: A wavy, festooned, feathery, or zigzag pattern of decoration in two or more colors, produced by applying threads of opaque glass of a color different from that of the molten glass body. The threads are rolled into the glass body by **marvering**, after which they are combed or dragged to achieve the desired effect.

Commedia dell'arte figures (Italian): Figures representing the 16 characters in Italian commedia dell'arte, a theatrical genre that was especially popular in the 17th and 18th centuries. **Flameworkers** made models of these figures, often copying illustrations in Lelio Riccoboni's *Histoire du théâtre italien* (1728-1731).

Compote: A dish, usually with a stem and a base, and sometimes with a cover, for serving compote (fruits cooked in syrup), or a smaller dish of similar form used for individual servings.

Concentric paperweight: A type of paperweight in which the slices of cane are arranged in concentric circles.

Condiment set: A group of matching vessels, usually with a tray or rack, that includes containers for salt, pepper, and mustard, and perhaps also **cruets** for oil and vinegar. See **Castor**.

Cone beaker: A drinking vessel in the form of an inverted cone. Cone beakers were used in several cultures, including ancient Rome, Sasanian Iran, and early medieval Europe.

Copper-wheel engraving: A technique of decorating the surface of an object. Copper disks (wheels) of various sizes and rim profiles are rotated on a spindle. An abrasive such as Carborundum® (in the past, emery was frequently used), mixed with oil, is applied to the edge of the wheel. The wheel presses the abrasive against the glass so that it removes the surface by grinding.

Cord: Accidental colorless streaks in glass caused by local differences in refractive indexes. Cord is often produced by poor mixing of the **batch**.

Core: The form to which molten glass is applied in order to make a **core-formed** vessel. In pre-Roman times, the core is thought to have been made of animal dung mixed with clay.

Core forming: The technique of forming a vessel by trailing or gathering molten glass around a **core** supported by a rod. After forming, the object is removed from the rod and annealed. After **annealing,** the core is removed by scraping.

Cowhorn: The large end of a **mosaic glass cane** that is shaped like the tapering horn of a cow.

Cracking off: The process of detaching a glass object from the **blowpipe** or **pontil**.

Crackle glass: See **Ice glass**.

Creamer: A small pitcher for serving cream.

Crimper: A tool used for decorating objects by giving them a crimped or wavy edge.

Cristallo (Italian, "crystal"): A term first used in Venice in the 14th century to describe glass that resembles colorless rock crystal. Most Venetian *cristallo*, however, has a gray or brownish tint.

Crizzling, crisseling: A chemical instability in glass caused by an imbalance in the ingredients of the **batch**, particularly an excess of **alkali** or a deficiency of stabilizer (usually **lime**). The instability of the glass results in an attack by atmospheric moisture, which produces a network of cracks in the surface. Crizzling can be slowed or perhaps even halted, but it cannot at present be reversed.

Crown glass: Sheet glass made by blowing a **parison**, cutting it open, and rotating it rapidly, with repeated reheating, until the centrifugal force has caused it to become a flat disk. After **annealing**, the disk is cut into panes of the required shape and size. "Bull's-eye" panes come from the centers of the disks and preserve the thickened area where the parison was attached to the **pontil**.

Crown weight: A hollow paperweight that incorporates thin white or colored filigree canes arranged vertically on the sides and drawn together at the top.

Making crown glass. From Diderot, Receuil de planches...,
about 1775.

Cruet: A small, ewerlike vessel, usually with a lip or
spout, a handle, and a stopper, for serving condiments at
the table.

Crystal: A popular term for colorless **lead glass** which
has a high refractive index and consequently is particularly
brilliant. Today, the word is often used to describe any
fine glass tableware.

Cullet: (1) Raw glass or pieces of broken glass from a
cooled **melt**; (2) scrap glass intended for recycling.

Cup plate: A small plate on which users set their tea or
coffee cup while drinking from the saucer. Between about
1825 and about 1865, it was fashionable to drink from
saucers, and cups were placed on cup plates to avoid
staining the tablecloth.

Cupping glass: A small cup in which a partial vacuum is created for cupping. Cupping is the technique of drawing blood to the surface of the body, usually for bloodletting.

Custard glass: A vessel for an individual serving of custard, a sweetened mixture of milk and eggs, which may be baked, boiled, or frozen.

Cutting: The technique whereby glass is removed from the surface of an object by grinding it with a rotating wheel made of stone, wood, or metal, and an abrasive suspended in liquid. See also **copper-wheel engraving**, **carving**, and **wheel engraving**.

Cylinder glass: Window glass made by inflating a large **gather** and swinging it until it forms a cylinder. The cylinder is then detached from the **blowpipe**, and both ends are removed with **shears**. Next, the cylinder is cut lengthwise, reheated, and either tooled or allowed to slump until it assumes the form of a flat sheet. After **annealing**, the sheet is cut into panes.

D

Daumenglas (German, "thumb glass"): A large cylindrical or barrel-shaped **forest glass** beaker with circular indentations for the user's fingers and thumbs. *Daumengläser* were made in Germany and the Netherlands in the 16th and 17th centuries.

Decanter: A decorative bottle with a stopper, used for serving wines and spirits.

Decolorizer: A substance (such as manganese dioxide or cerium oxide) used to remove or offset the greenish or brownish color in glass that results from (1) iron impurities in the **batch** or (2) iron or other impurities in the **pot** or elsewhere in the production process.

Daumenglas. *Germany, 17th century. OH. 36.8 cm.*

Depression glass: Inexpensive, machine-pressed American glassware made between about 1920 and 1950.

Devitrification: (1) The process whereby glass becomes partly crystallized as it cools (usually too slowly) from the molten state; (2) the crystals formed by this process. Devitrification may also occur on the surface as a result of unsuccessful **annealing** or accidental heating to a high temperature. It is *not* caused by chemical reaction between glass and its environment, which is known as **weathering**.

Diamond air trap: Decoration consisting of bubbles of air trapped in the glass in a diamond-shaped pattern. This is achieved by blowing a **gather** of glass into a mold with projections of the desired design, withdrawing it, and covering it with a second gather, which traps pockets of air in the indentations. This technique was patented by W. H., B. & J. Richardson of England in 1857.

Diamond-point engraving. France or the Netherlands, 17th century. D. 48.8 cm.

Diamond-point engraving: The technique of decorating glass by scratching the surface with a diamond, introduced by the Venetians in the 16th century and carried to some of its greatest artistic heights in the Netherlands during the 17th century. See **Stippling**.

Diatreta: A term used by Frederick Carder (1863-1963) to describe **openwork** objects, which he made by **lost wax casting**.

Diatretum, vas diatretum (Latin, "openwork vessel"): A term frequently used to refer to a **cage cup**. The plural form is *vasa diatreta*.

Dichroic glass: Glass that is one color when seen by reflected light and another color when light shines through it. This is sometimes due to the presence of minute quantities of colloidal gold.

Die sinker: A maker of metal molds.

Dip mold: A cylindrical, one-piece mold that is open at the top so that the **gather** can be dipped into it and then inflated. See also **Optic mold**.

Dolphin candlestick: A **pressed glass** candlestick with a stem in the form of a dolphin, originally made in New England between about 1840 and 1860.

Double cruet: See **Gemel**.

Double-walled: See *Zwischengoldglas*.

Dragon-stem goblet: A type of **goblet** with the stem in the form of a dragon with a convoluted body, outspread wings, open jaws, and a crest. Known in Italian as *vetri a serpenti* ("serpent glasses"), dragon-stem goblets were first made in Venice in the 17th century. They were imitated in the Netherlands by producers of *façon de Venise* glass.

Drawn stem: The stem of a drinking or serving vessel that is drawn out from the main **gather** rather than formed from a separate gather and then applied.

Dresser set: A set of matching perfume bottles, powder jars, and similar containers, kept on a woman's dresser.

Drip pan: A fixed or removable tray beneath the socket of a candlestick or the **font** of a lamp, which prevents spilled fuel or molten wax from escaping.

Dromedary flask: A small container consisting of a blown glass flask adorned with trails that may form an openwork "cage" and fused to the back of a camel- like animal, made by manipulating hot glass. Dromedary flasks were made in Syria between about the sixth and eighth centuries A.D.

Drop burner: A burner that was dropped into the **font** of a whale oil lamp and held above the fuel by a metal plate larger than the aperture in the font.

Enameled bottle and tumblers. Venice, about 1730-1735. H. (bottle) 27.5 cm.

E

Eglomisé: See **Verre églomisé**.

Egyptian blue: A synthetic material, copper calcium tetrasilicate, with a distinctive blue color. In antiquity, Egyptian blue was made by heating together **silica**, **lime**, and a copper-containing ingredient. It is often confused with **faience** and misleadingly called **frit**.

Enamel: A vitreous substance made of finely powdered glass colored with metallic oxide and suspended in an oily medium for ease of application with a brush. The medium burns away during firing in a low-temperature **muffle kiln** (about 965°-1300° F or 500°-700° C). Sometimes, several firings are required to fuse the different colors of an elaborately enameled object.

Encased glass: An object, such as a paperweight, that is covered with a layer of colorless glass.

Engraving: The process of cutting into the surface of an **annealed** glass object either by holding it against a rotating copper wheel fed with an abrasive or by scratching it, usually with a diamond. See also **carving**, **cutting**, and **stippling**.

Epergne (French): A composite, frequently tiered centerpiece used on the dinner table for serving or display in the late 18th and 19th centuries.

Eye bead: A bead decorated with applied or embedded circular elements that resemble eyes.

F

Faceting: The process of grinding and polishing an object to give the surface a pattern of planes or facets.

Façon de Venise (French, "style of Venice"): Glass made in imitation of Venetian products, at centers other than Venice itself. *Façon de Venise* glass was popular in many parts of Europe in the 16th and 17th centuries.

Faience: A fired silica body containing small amounts of **alkali**, and varying greatly in hardness depending on the degree of **sintering**. It is covered with glaze, which may also be present interstitially among the quartz grains within the body. The term "glassy faience" is often used to describe a faience in which the reactions have proceeded to such an extent that the glass phase defines the visual appearance of the material.

Fake: A genuine object that has been altered or "improved" for the purpose of enhancing its value.

Favrile glass: A type of glass with an **iridescent** surface, patented by Louis Comfort Tiffany (1848-1933) in 1894.

Figural bottle: A molded bottle in the form of a head or any of a variety of objects. The term is usually applied to

American glass of the late 19th and early 20th centuries.

Filigrana, vetro a filigrana (Italian, "filigree glass"): The generic name for blown glass made with colorless, white, and sometimes colored canes. The *filigrana* style originated at Murano in the 16th century and spread rapidly to other parts of Europe, where ***façon de Venise*** glass was produced. Manufacture at Murano continued until the 18th century, and was revived in the 20th century. For the main types of ***filigrana***, see ***Vetro a fili***, ***vetro a reticello***, and ***vetro a retorti***.

Finger bowl: A bowl to hold water for rinsing the fingers at the table.

Finial: An ornamental knob.

Finishing: The process of completing the forming or decoration of an object. Finishing may take the form of manipulating the object into its final shape while it is hot, of **cracking off** prior to **annealing**, or of grinding, **cutting**, or **polishing**.

Fire clay: Clay capable of being subjected to a high temperature without fusing, and therefore used for making crucibles in which glass **batches** are melted. Fire clay is rich **silica**, but contains only small amounts of **lime**, iron, and **alkali**.

Fire polishing: The reintroduction of a vessel into the **glory hole** to melt the surface and eliminate superficial irregularities.

Firing: The process of (1) heating the **batch** in order to fuse it into glass by exposing it to the required temperature in a crucible or **pot**, (2) reheating unfinished glassware while it is being worked, or (3) reheating glassware in a **muffle** to fuse **enamel** or **gilding**. The melting of the

Flameworked figurine.
France, probably Nevers,
early 18th century.
H. 5.0 cm.

batch may require a temperature of about 2575-2040° F
(1300°-1500° C), whereas the **muffle kiln** may require a
temperature of only about 960°-1320° F (500°-700° C).

Firing glass: A drinking glass with a bowl, a short stem,
and a thick foot. On ceremonial occasions, firing glasses
were rapped loudly on the table, making a noise that
resembled a volley of gunfire.

Flameworking: The technique of forming objects from
rods and tubes of glass that, when heated in a flame,
become soft and can be manipulated into the desired
shape. Formerly, the source of the flame was an oil or
paraffin lamp used in conjunction with foot-powered
bellows; today, gas-fueled torches are used.

Flashing: The application of a very thin layer of glass of
one color over a layer of contrasting color. This is
achieved by dipping a **gather** of hot glass into a crucible
containing hot glass of the second color. The upper layer

may be too thin to be worked in relief. "Flashing" is sometimes used (erroneously) as a synonym for **casing**.

Flat bouquet: A decorative arrangement of **canes** in a **paperweight**. The canes imitate a bouquet of flowers and leaves, the flat top of which is parallel to the bottom of the paperweight.

Flint glass: A misnomer for English and American **lead glass**. The term came into use in the 17th century, when ground or calcined flint temporarily replaced **sand** as the source of **silica** for glassmaking in England.

Flügelglas (German, "wing glass"): See **Winged goblet**.

Flute: A very tall and slender wineglass on a short stem.

Fluting: Decoration that consists of narrow vertical grooves (flutes).

Flux: A substance that facilitates fusion. For example, a flux is added to the **batch** in order to facilitate the fusing of the **silica**. Fluxes are also added to **enamels** in order to lower their fusion point to below that of the glass body to which they are to be applied. **Potash** and **soda** are fluxes.

Folded rim: A rim that has been folded to double its thickness and thereby increase its strength.

Fondo d'oro (Italian, "base of gold"): See **Gold glass**.

Font: The reservoir for oil in a lamp.

Foot-ring: A separate ring of glass added to the base after the body of the vessel is formed.

Forest glass: Glass made in the rural glasshouses of central and northern Europe in the late Middle Ages and

Forest glass beaker. Germany, 15th-16th century. H. 7.8 cm.

the early modern period. Most forest glass was fluxed with **potash** derived from the wood with which the **furnaces** were fueled. It is green because of iron impurities in the sand from which it was made. The German term for forest glass is ***Waldglas***.

Forgery: A copy or imitation of an object, made with the intention of deceiving prospective owners into believing that it is the genuine article.

Former mold: A mold with the same shape as the desired object, usually a vessel. Flat glass **blanks** are made into vessels by **sagging** them over or into former molds.

Founding: The initial phase of melting **batch**. For many modern glasses, the materials must be heated to a temperature of about 2450° F (1400° C). This is followed by a maturing period, during which the molten glass cools to a working temperature of about 2000° F (1100° C).

Free-blown (off-hand blown) glass: Glassware shaped solely by inflation with a **blowpipe** and manipulation with **tools**.

Frigger: An object made by a glassworker on his own time. Most friggers were made from the molten glass that remained in the **pot** at the end of the day. Such glass was considered to be a worker's perquisite.

Frit, fritting: Batch ingredients such as **sand** and **alkali**, which have been partially reacted by heating but not completely melted. After cooling, frit is ground to a powder and melted. Fritting (or **sintering**) is the process of making frit.

Frosting: (1) A **matte finish** produced by exposing the object to fumes of **hydrofluoric acid**; (2) a network of small surface cracks caused by **weathering**.

Furnace: An enclosed structure for the production and application of heat. In glassmaking, furnaces are used for melting the **batch**, maintaining **pots** of glass in a molten state, and reheating partly formed objects at the **glory hole**.

Fusing: (1) The process of **founding** or melting the **batch**; (2) heating pieces of glass in a **kiln** or **furnace** until they bond (see **casting** and **kiln forming**); (3) heating **enameled** glasses until the enamel bonds with the surface of the object.

G

Gadget: A metal rod with a spring clip that grips the foot of a vessel and so avoids the use of a **pontil**. Gadgets were first used in the late 18th century.

Gilded bottle. England, probably Bristol, about 1790. H. 13.1 cm.

Gadroon: A flutelike decorative motif, usually short in proportion to its width, that often approaches an oval form.

Gaffer (English, corruption of "grandfather"): The master craftsman in charge of a **chair**, or team, of hot-glass workers.

Gather: (Noun) A mass of molten glass (sometimes called a gob) collected on the end of a **blowpipe**, **pontil**, or **gathering iron**; (verb) to collect molten glass on the end of a tool.

Gathering iron: A long, thin rod used to **gather** molten glass.

Gemel: A pair of bottles blown separately and then fused, usually with the two necks pointing in different directions.

Gilding: The process of decorating glass by the use of gold leaf, gold paint, or gold dust. The gilding may be applied with **size**, or amalgamated with mercury.

Glass: A homogeneous material with a random, liquidlike (non-crystalline) molecular structure. The manufacturing process requires that the raw materials be heated to a temperature sufficient to produce a completely fused **melt**, which, when cooled rapidly, becomes rigid without crystallizing.

Glazier: A craftsman who paints and/or assembles glass windows.

Glory hole: A hole in the side of a glass **furnace**, used when reheating glass that is being fashioned or decorated. The glory hole is also used to **fire-polish cast** glass to remove imperfections remaining from the **mold**.

Goblet: A drinking vessel with a bowl that rests on a stemmed foot.

Gold-glass roundel. Roman Empire, Italy, 4th century A.D. D. 9.7 cm.

Gold glass: The term applied to several types of Hellenistic and ancient Roman glass objects decorated with designs cut and/or engraved in gold leaf, which is sandwiched between two fused layers of glass. Hellenistic gold glass was made by sandwiching the decoration between two closely fitting, cast, ground, and polished vessels, which were then fused. Many Roman gold glasses apparently were made by applying the gold leaf to the surface of an object, reheating it, and inflating a **parison** against the decorated surface.

Gold ruby: Deep red glass colored by the addition of gold chloride to the **batch**. The method of making gold ruby glass was perfected by Johann Kunckel (about 1630-1703) in Potsdam shortly before 1679.

Gold sandwich glass: See **Gold glass**.

Gold-band mosaic glass: A variety of **ribbon glass**, which includes **canes** composed of bands of gold foil laminated between two layers of colorless glass. Gold-band mosaic glass was made in parts of the Roman world in the first century B.C. and the first century A.D.

Graal glass: A type of decorative glass developed by Orrefors of Sweden in 1916. The design is **carved**, **engraved**, or etched on a **parison** of colored glass, which is then reheated, encased in a thick outer layer of transparent glass of a different color, and inflated.

Grape flask: An ancient Roman **mold-blown** flask with the body in the form of a bunch of grapes.

Grenade: A type of bottle with a short, narrow neck and a globular body, which apparently was filled with water and thrown into flames to serve as a fire extinguisher. **Aeolipiles** are sometimes identified, probably incorrectly, as grenades.

Grisaille (French *gris*, "gray"): (1) A method of decorative painting in monochrome gray especially, but not

Head flask. Roman Empire, eastern Mediterranean, 4th century A.D. H. 19.6 cm.

exclusively, on **stained glass** windows; (2) brown paint made from iron oxide, which, when fused to the glass, defines details in a stained glass window.

Grozing: The process of breaking away the edge of a glass object with a grozing iron or pliers in order to shape it.

H

Hand cooler: A solid, egg-shaped piece of glass or decorative stone, said to have been used to cool the palms of a woman's hands.

Hand press: A tool shaped like a pair of pliers, with flat jaws containing molds. Hand presses were used extensively in Europe for making chandelier parts. Later, they were introduced in the United States for pressing stoppers and bases.

Head flask: A **mold-blown** flask with the body in the form of a human head. Head flasks were popular in the Roman Empire, and examples were made from the first to fourth centuries A.D. Vessels decorated with two faces placed back to back are sometimes known as "janiform" head flasks (from Janus, the spirit of doorways, who was represented as a double-faced head).

Hedwig beaker: A very rare type of thick-walled glass beaker with **relief-cut** decoration of lions, griffins, eagles, and other motifs. Hedwig beakers date from about the 12th century, but their place of manufacture is unknown. They are so called because one of the surviving examples is said to have belonged to Saint Hedwig of Silesia (d. 1243).

Hinterglasmalerei (German, "painting behind glass"): See **Reverse painting**.

Hochschnitt (German, "high cut"): See **Relief cutting**.

Hofkellereiglas (German, "court wine-cellar glass"): A drinking glass used in the buttery of a German court.

Hookah (Arabic *huqqa*): A bell-shaped or globular bottle that is part of the water pipe used in the Islamic world and India for smoking tobacco. The smoke passes through the water-filled bottle before the smoker inhales it.

Hot-formed, hot-worked: The generic term for glass that is manipulated while it is hot.

Humidor: A closed container in which the air is kept appropriately humidified (for example, for keeping cigars).

Humpen (German, "beaker"): A large, cylindrical beaker, usually with enameled decoration, made in Germany, Bohemia, and Silesia between the 16th and 18th centuries, and used mainly for drinking beer.

Hyalith glass: Two varieties of glass, opaque black and opaque red, developed by the Bohemian glassmaker Jiří von Buquoy (1781-1851) and patented in 1817 and 1819, respectively.

Hydrofluoric acid: A highly corrosive acid that attacks silicates such as glass. Pure hydrofluoric acid dissolves glass, leaving a brilliant, **acid-polished** surface.

I

Ice glass: A decorative effect that causes the surface of the glass to resemble cracked ice. This is achieved by plunging a **parison** of hot glass into cold water and withdrawing it quickly. The thermal shock creates fissures in the surface, and these impart a frosted appearance after the parison has been reheated to allow the forming process to continue.

Inclusions: A collective term for bubbles, metal and glass particles, and other foreign materials that have been added to the glass for decorative effects.

Incrustation: See **Sulphide**.

Inlay: Any object embedded in the surface of a larger object. See also **Marquetry**.

Intaglio (Italian, "engraving"): A method of engraving whereby the ornamentation is cut into the object and lies below the surface plane. The German name for this technique is *Tiefschnitt*.

Intarsia glass (from Italian *intarsiatura*, "marquetry"): A type of glass developed by Frederick Carder (1863-1963) about 1920. A design of colored glass was applied to a **parison** of a different color, then **flashed** with a second parison of the same color as the first.

Iridescence. Jack-in-the-pulpit vase. U.S., Tiffany Studios, 1912. H. 47.6 cm.

Intercalaire (French, "inserted"): The process of applying two layers of decoration, the first being covered with a skin of glass that serves as the surface for the second.

Iridescence: The rainbowlike effect that changes according to the angle from which it is viewed or the angle of incidence of the source of light. On ancient glass, iridescence is caused by interference effects of light reflected from several layers of **weathering** products. On certain 19th- and 20th-century glasses, iridescence is a deliberate effect achieved by the introduction of metallic substances into the **batch** or by spraying the surface with stannous chloride or lead chloride and reheating it in a reducing atmosphere.

J

Jacks: A tool with two metal arms joined at one end by a handle. The distance between the arms is controlled by the glassworker, who uses jacks, or **pucellas**, to form the mouths of open vessels.

Jacobite glass: A 17th-century English drinking vessel used for toasting Prince Charles Edward Stuart ("Bonnie Prince Charlie"). The Jacobites were supporters of the exiled King James II, who abdicated in 1698, and of his descendents James Edward Stuart (the "Old Pretender") and his son Charles Edward Stuart (the "Young Pretender"). Before the defeat of the Young Pretender in 1746, Jacobite glasses were usually engraved with the English rose, representing the Crown, and an optimistic motto such as *Redeat* (Latin, "May he return"). After 1746, glasses at first bore cryptic symbols and messages, but later secrecy was abandoned. See also **Williamite glass**.

Jelly glass: A vessel for serving jelly and other desserts. Jelly glasses usually have an inverted conical bowl, a square stem, and a foot. They may have one or two handles.

Jugendstil (German, "youth style"): See **Art Nouveau**.

K

Kalkglas (German, "chalk glass"): See **Chalk glass**.

Kantharos (Greek), **cantharus** (Latin): A drinking vessel with a bell-shaped body, a foot, and two handles.

Kick: A concavity in the base of a vessel, where it has been pushed in by a tool. The provision of a kick strengthens the bottom of the vessel and reduces its capacity.

Kiln: An oven used to process a substance by burning, drying, or heating. In contemporary glassworking, kilns are used to **fuse enamel** and for **kiln forming** processes such as **slumping**.

Kiln forming: The process of **fusing** or shaping glass (usually in or over a **mold**) by heating it in a **kiln**.

Kohl tube. Achaemenian Empire, Iran, 5th-4th century B.C. H. 8.8 cm.

Knop: A component, usually spherical or oblate, of the stem of a drinking glass, hollow or solid, used either singly or in groups, and placed contiguously or with intermediate spacing; also the **finial** at the center of a lid.

Kohl tube (Arabic *kuhl*, a cosmetic preparation): A small, tubular container for cosmetic preparations such as kohl. Kohl is a black powder, usually prepared from antimony, that is used in many parts of the Islamic world to darken the eyelids.

Krateriskos (Greek, "small mixing bowl"): A small vessel with a wide mouth and body, and a foot. The term is often used to describe certain **core-formed** Egyptian vessels of the second millennium B.C.

Krautstrunk (German, "cabbage stalk"): A type of beaker with a cup-shaped mouth and a cylindrical or barrel-shaped body decorated with **prunts**, made in Germany between the 15th and 17th centuries. It was the forerunner of the *Römer*.

Kunckel red: See **Gold ruby**.

Kurfürstenhumpen (German, "electors beaker"): A
Humpen decorated with images of the Holy Roman
Emperor and the seven Electors of the empire.

Kuttrolf (German): A flask with the neck divided into two
or more tubes. The *Kuttrolf*, which has Roman
antecedents, was produced by German glassworkers in the
later Middle Ages; it is also found among Venetian and
façon de Venise glasses of the 16th and 17th centuries.

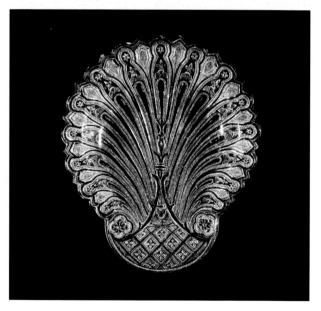

Lacy-pattern tray. U.S., New England, about 1830-1845.
L. 23.8 cm.

L

Lacy-pattern glass: Nineteenth-century **pressed glass**
whose patterns include extensive **stippling** to produce a
bright, lacelike effect that conceals wrinkles caused when
the cold plunger of the pressing machine came into contact
with the hot glass.

46

Lagynos (Greek): A pitcher with a tall, narrow neck and a wide body.

Lampworking: See **Flameworking**.

Lathe cutting: The technique whereby a **blank** in the general shape of the finished object is mounted on a lathe and (in antiquity) turned with the aid of a bow or handled wheel, while a tool fed with abrasive is held against the surface in order to polish it, modify the profile, or cut it.

Latticino, latticinio (from Italian *latte*, "milk"): A term formerly used to describe *filigrana* glass. It has now been abandoned.

Lattimo (from Italian *latte*, "milk"): Opaque white glass, usually opacified by tin oxide or arsenic.

Laub- und Bandelwerk (German, "leaf and strapwork"): A type of interlaced ornament consisting of foliage and strapwork, popular in Germany and Bohemia in the 18th century.

Lead glass: Glass that contains a high percentage of lead oxide (at least 20 percent of the **batch**). In modern times, glass of this type was first used by George Ravenscroft (1632-1683) about 1676. It is relatively soft, and its refractive index gives a brilliance that may be exploited by covering the surface with polished wheel-cut facets.

Lehr, leer: The oven used for annealing glassware. Early lehrs were connected to the **furnace** by flues, but the difficulty in controlling heat and smoke made the arrangement impracticable. Later lehrs were long, brick-lined, separately heated tunnels through which the glass objects were slowly pushed; the glass remained in the lehr for several hours, while it was gradually reheated and then uniformly cooled. Today, lehrs work on a conveyor belt system.

Lily-pad decoration. U.S., about 1835-1850.
OH. (bowl) 27.2 cm.

Lentoid flask: A flask with a lens-shaped body.

Lily-pad decoration: Decoration consisting of a **gather** around the base of the vessel, which has been drawn upward in four or more projections with rounded ends. Lily-pad decoration was introduced to America by German glassworkers. It became popular in New England, New York, and New Jersey in the second quarter of the 19th century.

Lime: Calcined limestone, which, added to the glass **batch** in small quantities, gives stability. Before the 17th century, when its beneficial effects became known, lime was introduced fortuitously as an impurity in the raw materials.

Linen smoother: An object believed to have served as a pressing iron. The earliest linen smoothers date from the Middle Ages, and the latest were made in the 18th century.

Lion-mask stem: A hollow stem made by blowing the **gather** into a **mold** patterned with two lion's masks, usually separated by festoons. Lion-mask stems, first used in

Venice in the 16th century, subsequently became one of the hallmarks of *façon de Venise* glass.

Lipper: A glassworker's tool made of wood in the shape of a cone and with a handle. It is used to form the lip at the mouth of a vessel.

Lithyalin (from Greek *lithos*, "stone"): A type of glass, developed in Bohemia by Friedrich Egermann (1777-1864), that is opaque and has a marbled surface resembling semiprecious stones.

Lost wax casting: A technique adapted from metalworking. The object to be fashioned in glass is modeled in wax and encased in clay or plaster that is heated. The wax melts and is released through vents or "gates," also made of wax, which have been attached to the object before heating; the clay or plaster dries and becomes rigid. This then serves as a mold, into which molten or powdered glass is introduced through the gates. If powdered glass is used, the mold is heated in order to fuse the contents. After **annealing**, the mold is removed from the object, which is then finished by grinding, **fire polishing**, or **acid etching**.

Lotus-bud beaker: A first-century A.D. Roman **mold-blown** vessel decorated with rows of oval or almond-shaped bulges. Although the bulges are usually described as lotus buds, they are probably derived from representations of knotholes in the club of the mythical hero Hercules.

Loving cup: A large drinking vessel with two or more handles, passed around at banquets and similar gatherings so that several persons could drink from it in turn.

Luster: (1) A shiny metallic effect made by painting the surface with metallic oxides that have been dissolved in acid and mixed with an oily medium. Firing in oxygen-free conditions at a temperature of about 1150° F (600° C) causes the metal to deposit in a thin film that, after cleaning, has a distinctive shiny surface. (2) A glass lighting device, such as a **candelabrum** or **candlestick**, decorated with pendent prismatic drops.

M

Maigelein (German): A type of small, hemispherical cup on a base with a **kick**, usually with vertical or swirled ribs, made in Germany in the 15th and 16th centuries.

Marbled glass: Glass decorated with streaks of two or more colors, resembling marble. Marbled glass was a Venetian specialty from the 15th to 17th centuries, but it was also made in other times and places.

Marquetry: A decorating technique whereby pieces of hot glass are applied to still molten glass and **marvered** into the surface, creating an inlaid effect. After the glass is cooled, it is possible to further emphasize these areas by **carving** and **engraving**. See also **Inlay**.

Martelé (French, "hammered"): The word used to describe the multifaceted, **wheel-engraved** surface favored by Emile Gallé (1846-1904), Daum, and others to create a textured background that resembles beaten metal.

Marver (French *marbre*, "marble"): A smooth, flat surface, over which softened glass is rolled in order to smooth it or to consolidate applied decoration.

Masonic glass: A glass object decorated with emblems or inscriptions associated with Freemasons.

Using the marver. From Diderot, Receuil de planches...,
about 1775.

Matsu-no-ke: A design developed by Frederick Carder
(1863-1963) and registered by Stevens & Williams of
England in 1884. Its distinctive feature is the presence of
applied and tooled sprays of blossoms influenced by
Japanese designs. Carder also used the design at Steuben
Glass Works in Corning, New York, in the 1920s.

Matte finish: A non-shiny finish made by grinding,
sandblasting, or exposing the surface to fumes of
hydrofluoric acid. See **Frosting**.

Melt: The fluid glass produced by melting a **batch** of raw
materials.

Mercury bottle: A type of ancient Roman **mold-blown**
bottle with a tall body of square or polygonal cross section,
the underside of which bears a representation in relief of
the god Mercury. Mercury, the messenger of the gods,
was associated with commerce.

Merese: A flattened, collarlike **knop** placed between the bowl and the stem, on the stem, or between the stem and the foot of a wineglass.

Metal: A term frequently used as a synonym for **glass**. It is misleading because glass is not a metallic substance, and its use is discouraged.

Metallic oxide: The oxide of a metal. Oxides may be used to color glass and **enamel**, or to produce lustered or iridized surfaces. The resultant color depends primarily on the oxide used, but it can be affected by the composition of the glass itself and the presence or absence of oxygen in the **furnace**. See **Iridescence** and **Luster**.

Mezza-forma (Italian, "half-mold"): A term applied to the process of making vertical ribs on the lower part of a blown glass object by inflating the **parison** in a **dip mold**.

Milchglas (German, "milk glass"): See *Lattimo*.

Milled threading: Decoration consisting of a **trail** (thread) that has been closely notched either by the use of a runner like a roulette or by repeated indentation with the edge of the **jacks**.

Millefiori (Italian, "1,000 flowers"): See **Mosaic glass**.

Mirror monogram: A monogram written in such a way that each letter is reversed to produce its mirror image, the letter and its image being combined to give a symmetrical ornamental form.

Moil: See **Overblow**.

Molar flask: A small flask standing on very short feet resembling the roots of a tooth. The molar flask is a characteristic type of ninth- to 14th- century Islamic

Mold. Islamic, western Asia, 11th-13th century. H. 11.4 cm.

perfume bottle. Many examples have simple wheel-cut decoration.

Mold: A form used for shaping and/or decorating molten glass. Some molds (e.g., **dip molds**) impart a pattern to the **parison**, which is then withdrawn, and blown and tooled to the desired shape and size; other molds are used to give the object its final form, with or without decoration.

Mold blowing: Inflating a **parison** of hot glass in a mold. The glass is forced against the inner surfaces of the mold and assumes its shape, together with any decoration that it bears.

Mold mark: See: **Seam mark**.

Mold pressing: Forcing hot glass into an open or multi-part mold by means of a plunger.

Mosaic: A surface decorated with many small, adjoining pieces of varicolored materials such as stone or glass.

Mosaic glass: Objects made from preformed elements placed in a mold and heated until they fuse. The term "mosaic glass" is preferable to "**millefiori**," except in the case of Venetian or *façon de Venise* glass.

Mosque lamp: An Islamic lamp shaped like an inverted bell, with three or more handles from which it was suspended by chains. Many mosque lamps have **gilded** and **enameled** decoration, which often includes inscriptions naming the donor and quoting verses from the Koran.

Moss Agate glass: A variety of **Art Glass** developed by John Northwood (1836-1902) and Frederick Carder (1863-1963) in England in the late 1880s. It was made by casing a **parison** of **soda-lime glass** with colorless **lead glass**, then covering it with powdered glass of several colors, and casing it again with lead glass. The object was shaped and reheated, after which cold water was injected into it, causing the soda-lime glass to crackle.

Muffle: A **fire-clay** box in which glass (or porcelain) objects are enclosed, when placed in the **muffle kiln**, to protect them from the flames and smoke while being subjected to low-temperature firing, especially in the process of firing **enamels** and **gilding** at temperatures of about 950-1320° F (500°-700° C).

Muffle kiln: A low-temperature kiln for refiring glass to fuse **enamel**, fix **gilding**, and produce **luster**. See **Kiln**.

Murrhine, *murrina, murrino* (from Latin *murra*, apparently a stone from which costly vessels [*vasa murrina*] were made): The English adjective "murrhine" and the Italian adjective *murrino* are sometimes applied to **mosaic glass** and similar objects. There is no evidence that confirms the popular view that ancient Roman *vasa murrina* were made of glass. When used as a noun, *murrina* refers to a slice of a complex **cane**, while a *murrino* is an insert of multicolored glass embedded in a glass object.

Neon lighting. U.S., Paul Seide, 1986. H. 48.4 cm.

N

Nappy: In the United States, a small serving dish.

Natron: Sodium sesquicarbonate, originally obtained from the Wadi el-Natrun, northwest of Cairo. It was commonly used by Roman glassmakers as the **alkali** constituent of **batch**.

Nécessaire (French, "necessity"): A traveling case containing a drinking glass and a knife, fork, and spoon.

Nef (Old French *nef*, "ship"): A table ornament in the form of a ship, with the hull formed by blowing and the rigging consisting of **trails**. Sometimes, there is a spout in the hull.

Neoclassical: A style of art, architecture, literature, music, etc., that is based on, or influenced by, classical styles, especially the styles of ancient Greece and Rome.

Neon, neon lighting: Neon is an inert gas which, like some other gases, has the properties of high electrical

*Nipt-diamond-waies.
England, George
Ravenscroft, 1676-1678.
H. 18.6 cm*

conductivity and strong light-emissive power. Such gases may be introduced into evacuated glass tubes. Under these conditions, an electrical discharge causes the gas to emit light. Different gases emit different colors; for example, neon emits red, and xenon emits blue. Regardless of the gases employed, lighting of this type is known as neon lighting.

Newel: Usually, the post at the head and the foot of a stair, supporting the handrail. In the 19th century, glass **finials** were sometimes used to adorn newel posts.

Nipt-diamond-waies: The technique of pincering adjacent vertical ribs to form a diamond pattern. "Nipt-diamond-waies" was the term used by the English glassmaker George Ravenscroft (1632-1683) in a 1677 advertisement for his new **lead glass**.

Nuppenbecher (German, "drop beaker"): A beaker decorated with large, droplike **prunts**, which may be drawn out into pointed projections.

O

Obsidian: A volcanic mineral that was the first form of natural glass used by humans. It is usually black, but it can also be very dark red or green; its splinters are often transparent or translucent.

Oenochoe (Greek, "wine pourer"): A pitcher with a trefoil mouth, used in ancient Greece to transfer wine from the mixing bowl to the cup. Between the sixth and third centuries B.C., miniature **core-formed** oenochoes were used as perfume bottles.

Omom (perhaps from Arabic *qumqum*, "sprinkler"): A sprinkler with a tall, narrow neck and an oblate spheroid body. Omoms were used in the Islamic world for sprinkling perfume.

Omphalos bowl (Greek *omphalos*, "navel"): A bowl with a hollow, raised boss (the "navel") at the center.

Opal glass: Glass that resembles an opal, being translucent and white, with a grayish or bluish tint.

Opalescent glass: (1) A type of late 19th-century **Art Glass**, made by covering a **gather** of colored glass with a layer of colorless glass containing bone ash and arsenic. The **parison** was inflated in a **mold** to produce raised decoration. When the parison was reheated, the raised areas became opalescent. (2) A type of glass resembling the iridescent gemstone opal, which was developed by Frederick Carder (1863-1963) at Steuben Glass Works in Corning, New York.

Openwork: Work that is perforated. Openwork in glass objects may be made by creating a network of **trails**, by **casting** (see **Diatreta**), or by **cutting** (see **Cage cup**).

Optic mold: An open mold with a patterned interior in which a **parison** of glass is inserted, then inflated to decorate the surface.

Optical glass: Glass of extreme purity and with well-defined optical properties, which was originally created for making lenses and prisms.

Opus sectile (Latin, "cut work"): Decoration on a wall or floor, made by fitting together flat elements of different shapes and colors. The ancient Romans sometimes used glass to make *opus sectile* ornament.

Overblow: A by-product of the **mold-blowing** process, this is the portion of the **parison** that remains outside the mold. The overblow, or moil, is usually removed by **cracking off**.

Overlay: A layer of glass that covers a layer of a different color, often as the result of **casing** or **flashing**.

P

Pallet: A glassworker's tool consisting of a square piece of wood or metal and a handle. It is used to flatten the bases of vessels.

Palm column flask: A tall, narrow, cylindrical vessel decorated at the top with stylized palm fronds. Flasks of this type were made by **core forming** in Egypt in the 18th-19th Dynasties (about 1400-1250 B.C.). They were used as **kohl tubes**.

Palm cup: A shallow drinking vessel with a round base that fits in the palm of the hand.

Pane: A piece of flat sheet glass used for glazing windows.

Paperweight. France, Cristallerie de Pantin, 1870-1880.
D. 11.5 cm.

Paperweight: A small, heavy object designed to hold down loose papers. The first glass paperweights were made in the 1840s in Venice and France, and their manufacture spread rapidly to other parts of Europe and the United States. Glass paperweights ceased to be fashionable in the early 20th century, but the craft of making them revived in the 1950s.

Parison (French, *paraison*): A **gather**, on the end of a **blowpipe**, which is already partly inflated.

Passglas (German, "pass glass"): A tall, cylindrical drinking vessel with trailed or enameled horizontal marks. The drinker was supposed to gulp only enough to reach the next horizontal mark, and then pass the glass to the next person. If he drank too much, he was required to reach the next mark, and so on.

Pâte de verre (French, "glass paste"): A material produced by grinding glass into a fine powder, adding a binder to create a paste, and adding a fluxing medium to facilitate

Peachblow glass. U.S., about 1886-1891. OH. 25.5 cm.

melting. The paste is brushed or tamped into a mold, dried, and fused by firing. After **annealing**, the object is removed from the mold and finished.

Patella cup (Latin *patella*, "small dish, kneecap"): A first-century A.D. Roman drinking cup with a double-convex profile and a **foot-ring**.

Pattern-molded glass: Glassware that has been blown into a mold whose interior has a raised pattern so that the object shows the pattern with a concavity on the inside, underlying the convexity on the outside. Pattern molds are not used to impart the final form to the object.

Peachblow glass: A type of **Art Glass** made by several American factories in the late 19th century. It resembled the peach bloom glaze on 17th- to 18th- century Chinese porcelain such as the celebrated Morgan Vase. Most Peachblow glass had a surface that shaded from opaque cream to pink or red, sometimes over opaque white. Similar glass was made in England by Thomas Webb & Sons and Stevens & Williams.

Pegging: The process of pricking molten glass with a tool that leaves small, air-filled hollows. When the glass is covered with a second **gather**, the hollows become **air traps**.

Phiale (Greek): A broad, flat bowl for drinking or pouring libations.

Pick-up decoration: A technique whereby a hot **parison** is rolled in chips of glass, which are picked up, **marvered**, and inflated.

Piece mold: A mold made of two or more parts.

Piggin: A small, cylindrical drinking vessel, often of wood, with one stave longer than the rest and serving as a handle.

Pilgrim flask: A flat flask with a ring on each side of the neck for the insertion of cords by which it may be suspended.

Pillar-molded glass: A term used by 19th-century English glassmakers to describe vessels with **mold-blown** vertical ribs but no corresponding indentations on the interior. This effect was achieved by partly inflating the **gather**, allowing it to cool sufficiently to become somewhat rigid, and then gathering an outer layer of glass around it. The **parison** was then further inflated in a ribbed **dip mold**, which shaped the soft outer layer without affecting the inner layer. The term is frequently but incorrectly applied to ancient Roman **ribbed bowls**, which were made in a different manner.

Pincers: A glassworker's tool used for decorating objects by pinching the glass while it is hot.

Plaque: An ornamental plate or tablet intended to be hung up as a wall decoration or inserted in a piece of furniture.

Plastic: Susceptible to being modeled or shaped. Glass is plastic when it is in a molten state.

Pokal (German): A covered **goblet** with a flared bowl, made mostly in Germany between the 17th and 19th centuries, and used for drinking toasts.

Polishing: Smoothing the surface of an object when it is cold by holding it against a rotating wheel fed with a fine abrasive.

Polycandelon (Greek, *polykandelon*): A lighting fixture consisting of a metal ring with apertures to hold cone-shaped lamps, suspended by three chains.

Pomona: A type of **Art Glass** developed by Joseph Locke at the New England Glass Company and patented in 1885. Made of colorless glass, it was **mold-blown** repeatedly, partly etched and stained amber or rose, and decorated with blue and amber garlands of flowers and fruits.

Pontil, pontil mark: The pontil, or punty, is a solid metal rod that is usually tipped with a wad of hot glass, then applied to the base of a vessel to hold it during manufacture. It often leaves an irregular or ring-shaped scar on the base when removed. This is called the "pontil mark."

Porringer: A shallow dish, usually with one or two horizontal handles, for porridge or similar food. Porridge is either a soft food made by boiling meal of cereals or legumes in milk or water until it becomes thick, or a soup of meat and vegetables, often thickened with cereal.

Porrón (Spanish): A type of drinking vessel with a narrow neck, a long, tapering spout, and no handle, used in Spain for drinking wine by pouring it into the mouth.

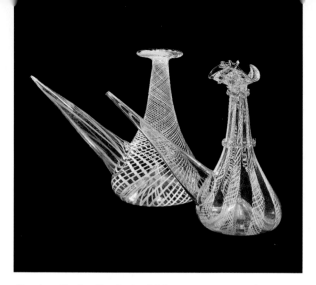

Porróns. Spain, Catalonia, 18th century.
H. (larger) 27.2 cm.

Posset pot: A spouted vessel used for consuming posset, a beverage of hot milk curdled by wine or ale, usually spiced or sweetened, and thickened with oatmeal or bread. The drinker sucked the liquid through the spout and ate the residue with a spoon.

Post technique: Instead of being applied to a vessel with a wad, the **pontil** is attached to a flat plate of glass called a "post," which is then affixed to the base or base-ring of the vessel.

Pot: A **fire clay** container in which the **batch** of glass ingredients is fused.

Potash: Potassium carbonate. It is an alternative to **soda** as a source of **alkali** in the manufacture of glass. Potash glass is slightly more dense than soda glass; it passes from the molten to the rigid state more quickly, and it is therefore more difficult to manipulate into elaborate forms. However, it is harder and more brilliant, and lends itself to decorative techniques such as facet cutting and **copper-wheel engraving**.

Potassium-lime glass: A form of glass containing three major compounds in varying proportions: **silica** (usually about 60-75 percent), **potash** (12-18 percent), and **lime** (5-12 percent). **Forest glass** is a common type of potassium-lime glass.

Preserving jar: A jar for preserving food. The term refers to jars with a variety of methods for sealing the contents. John Landis Mason perfected the first inexpensive method of sealing glass jars in 1858.

Pressed glass: Glassware formed by placing a blob of molten glass in a metal mold and then pressing it with a metal plunger or "follower" to form the inside shape. The resultant piece, termed "mold-pressed," has an interior form independent of the exterior, in contrast to **mold-blown** glass, whose interior corresponds to the outer form. The process of pressing glass was first mechanized in the United States between 1820 and 1830.

Prince Rupert's drop: A hollow glass object, about two inches long, with a bulbous end and a narrow, curving "tail." It is made by dropping a blob of hot glass into cold water and leaving it there until it has cooled. The rounded end resists a blow, but because of internal stress, the tail shatters into numerous fragments if it is broken or scratched. These objects, which have aroused great curiosity, were introduced into England by Prince Rupert (1619-1682), nephew of Charles II. Samuel Pepys described them in his diary on January 13, 1662.

Printy, printie: A circular or oval wheel-cut depression.

Prismatic cutting: A decorative pattern of long, mitered grooves, cut horizontally in straight lines so that the top edges of each groove touch the edges of the adjoining grooves. Prismatic cutting is usually found on the necks of pitchers and decanters.

Punch bowl, with ladle, tray, and glasses. France, Baccarat, 1867. H. 56.5 cm.

Prunt: A blob of glass applied to a glass object as decoration, but also to afford a firm grip in the absence of a handle.

Pucellas: See **Jacks**.

Punch bowl: A bowl for mixing and serving punch. Traditionally, punch was composed of wines or liquors, mixed with hot water or milk, and flavored with sugar, lemons, or spices. Fruit punches, with or without alcohol, are also common.

Punty: See **Pontil**.

Puzzle glass: See **Trick glass**.

Pyxis (Greek and Latin, pl. *pyxides*): A covered box for the toilet table, used to contain cosmetics, medicines, and jewelry.

Q

Quarry: A small, square- or diamond-shaped **pane**.

Quincunx (Latin, "five-twelfths"): An arrangement of five objects in a square or rectangle, with one at each corner and one in the middle, like the five spots on dice. **Prunts** and other motifs are sometimes arranged in a quincunx pattern.

R

Raised diamond cutting: An allover pattern of raised four-sided diamonds of pyramidal form, each with a sharp apex, cut with a mitered wheel. It was produced by English and Irish glass cutters between about 1780 and 1825.

Raspberry prunt: A flat, circular **prunt** with an impressed design resembling a raspberry.

Reactive glass: A type of glass, made by Louis Comfort Tiffany (1848-1933), that changed color when it was reheated.

Reducing atmosphere: An atmosphere in a **kiln** or **furnace** that is deficient in oxygen. Sometimes, a reducing atmosphere is created deliberately to reduce oxides to their metallic state, as in the case of **luster** pigments.

Refractory: A substance, usually clay, capable of resisting high temperatures.

Reichsadler Humpen (German, "imperial eagle beaker"): A ***Humpen*** decorated with a heraldic two-headed eagle whose wings bear the insignia of the Holy Roman Empire.

Relief cutting: A type of cut glass with decoration in high relief, made by removing the background.

Reichsadler Humpen.
Bohemia, 1574.
H. 26.4 cm.

Reticella: See *Vetro a reticella*.

Retort: A vessel with a long neck, bent downward, in which liquids subjected to distillation are heated.

Retorti, retortoli: See *Vetro a retorti*.

Reverse foil engraving: A decorative technique in which gold or silver leaf is applied to the back side of a piece of glass, engraved, and protected by varnish, metal foil, or another piece of glass. See also **Verre églomisé**.

Reverse painting: The term applied to a number of decorative techniques, all of which involve painting on the back side of the glass a design that is viewed from the front (that is, through the glass). Because of this, the painter must apply the pigments in the reverse of the normal order, beginning with the highlights and ending with the background.

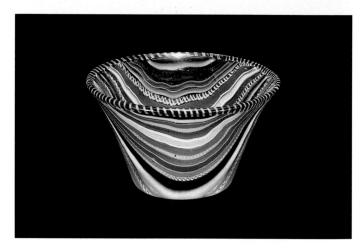

Ribbon glass cup. Roman Empire, 1st century B.C.-1st century A.D. H. 4.8 cm.

Rhyton (Greek): A drinking vessel used in many parts of the ancient world. Most ancient glass rhytons are of the Roman period. They are in the form of a horn, sometimes with the head of an animal at the tip. The tip is perforated, and one drank from it by holding the rhyton above one's head and catching the stream of liquid in the mouth.

Ribbed bowl: An ancient Roman vessel decorated with a continuous band of vertical ribs.

Ribbon glass: (1) A type of first-century A.D. Roman **mosaic glass** that consists of ribbonlike **canes** arranged in parallel rows or geometric patterns; (2) a type of *vetro a reticello* made in Venice and at other places where *façon de Venise* glass was produced.

Rigaree: A raised band or pattern of bands, usually made by **crimping** applied **trails**.

Rock crystal: Natural quartz, chemically pure silica. It is usually colorless (or nearly so). From earliest times, glassmakers sought to imitate it.

Rock-crystal engraving: A style of **copper-wheel engraving** that, combined with polishing, gave glass objects the appearance of **rock crystal**. The style was introduced by Thomas Webb & Sons of England in the 1870s.

Rod: A monochrome segment of glass cut from a **trail**.

Rod forming: The technique of winding molten glass around the tip of a narrow metal tool or wire coated with clay or "kiln wash" to act as a separating agent. It is used for making narrow vessels such as **kohl tubes**, and beads and pendants.

Rolling pin: A cylindrical object for rolling out dough or pasta to the required thickness. Many glass rolling pins are **friggers**, with a knob at each end so that they can be suspended by a cord.

Römer (German), ***roemer*** (Dutch): A drinking vessel with an ovoid mouth, a cylindrical body, and a conical foot. The body is usually decorated with **prunts**.

Rotary polishing: The process of polishing an object with tools and an abrasive, while turning it on a lathe.

Rummer: A type of 19th-century English **goblet**, with a short stem and a square or domed foot.

Rüsselbecher (German, "trunk beaker"): See **Claw beaker**.

S

Sagging: The process of reheating a **blank** so that it gradually flows under its own weight over or into a **former mold** and eventually assumes the shape of the mold.

Salt, saltcellar (French *salière*, "salt dish"): A small bowl used at the table for salt.

Salver: A tray for serving. At first, salvers were used primarily for presenting objects to rulers; more recently, the word is used to denote trays used for presenting letters or visiting cards, or for serving refreshments.

Sand: The most common form of **silica** used in making glass. It is collected from the seashore or, preferably, from deposits that have fewer impurities. For most present-day glassmaking, sand must have a low iron content. Before being used in a **batch**, it is thoroughly washed, heated to remove carbonaceous matter, and screened to obtain uniformly small grains.

Sand casting, sand molding: A glassforming technique in which molten glass is poured or ladled into a mold of compacted sand. A roughly textured granular surface results where the glass comes into contact with the sand.

Sand-core technique: A misnomer for **core forming**.

Sandblasting: The process of removing glass or imparting a **matte finish** by bombardment with fine grains of sand that is propelled by compressed air.

Satin glass: A 19th-century term for glass with a **matte finish**.

Scale: An accidental inclusion in glass, consisting of corrosion products detached from the metal implements used to stir the **batch** or to form the object.

Scarab (Latin *scarabaeus*, "beetle"): (1) A beetle, usually the scarabaeid beetle, which was revered by the ancient Egyptians; hence, (2) a gem in the form of a beetle, with a design in **intaglio** on the flat underside.

Schwarzlot *tumbler. Germany, early 18th century. H. 9.5 cm.*

Scheuer (German): A type of drinking glass with a short, cylindrical neck, a hemispherical body, and a single handle that projects outward and upward from the wall.

Schmelzglas (German, "enamel glass"): A term applied to several types of decorative glassware, including ***calcedonio*** and opaque white glass with a red overlay applied by **flashing**. It does not refer to glass decorated with **enamel**.

Schwarzlot (German, "black lead"): A sepia enamel first used in painting **stained glass** and later applied to glass vessels, either by itself or in combination with other **enamels** or gold.

Seal: Its many meanings include an emblem impressed on wax or some other **plastic** substance as evidence of ownership or authenticity. Since the 17th century, many bottles have borne stamped glass seals that identify the producer of the contents, the tavern in which they were used, or the individual for whom the contents were bottled.

Part of service. Austria, about 1916. H. (tallest) 32.8 cm.

Seam mark: A slight, narrow ridge on a glass object, which indicates that it has been made by molding. The seams appear where gaps in the joins between parts of the mold have permitted molten glass to seep during formation. On well-made pieces, the seam marks are usually smoothed away by grinding or **fire polishing**.

Seeds: Minute bubbles of gas, usually occurring in groups.

Service: A matching set of tableware.

Shearings (cuttings, clippings): Slivers of waste glass formed by trimming glassware during manufacture.

Shears: A tool used to trim excess hot glass from an object in the course of production.

Shot glass: A small drinking glass used to serve a single measure (shot) of liquor, usually whiskey.

Sidonian glass: A popular generic name for numerous first-century A.D. Roman mold-blown vessels. It is not known how many of these objects were actually made at Sidon, a city on the coast of Lebanon.

Silica: Silicon dioxide, a mineral that is one of the essential ingredients of glass. The most common form of silica used in glassmaking has always been **sand**.

Silver stain: A deep yellow stain made by painting the surface of the glass with silver sulfide and firing it at a relatively low temperature.

Silvered glass: A type of 19th-century glassware with an allover silver appearance, made by applying a solution of silver nitrate between the walls of a **double-walled** vessel. The solution was introduced through a hole in the base, which was then sealed to prevent the silver from oxidizing. Silvered glass is sometimes known, mistakenly, as "mercury glass."

Sintering: The process of heating a mixture of materials so that they become a coherent mass, but not melting them. See also **Frit**.

Size: In glassworking, the name applied to several glutinous materials, such as glue and resin, used to affix color or gold leaf.

Skyphos (Greek), **scyphus** (Latin): A cup with a foot and two opposed handles.

Slag glass: See **Marbled glass**.

Smalt: Colored glass, often deep blue glass colored with cobalt oxide. Smalts are finely ground to use as colorants for glass and enamel.

Snake-thread decoration: A type of decoration that

consists of **trails** applied in sinuous, snakelike patterns. It was made by the Romans between the second and fourth centuries A.D.

Snuff bottle, snuff box: A small bottle (in China) or box (in Europe) for powdered tobacco, or snuff. The habit of inhaling snuff, which spread to Europe from the Americas in the 17th century, was introduced to China in the 18th century.

Soda: Sodium carbonate. Soda (or alternatively **potash**) is commonly used as the **alkali** ingredient of glass. It serves as a **flux** to reduce the fusion point of the **silica** when the **batch** is melted.

Soda-lime glass: Historically, the most common form of glass. It contains three major compounds in varying proportions, but usually **silica** (about 60-75 percent), **soda** (12-18 percent), and **lime** (5-12 percent). Soda-lime glasses are relatively light, and upon heating, they remain **plastic** and workable over a wide range of temperatures. They lend themselves, therefore, to elaborate manipulative techniques.

Soffietta (Italian): A tool used as a puffer to further inflate a vessel after it has been removed from the **blowpipe** and is attached to the **pontil**. It consists of a curved metal tube attached to a conical nozzle. The glassblower reheats the vessel and inserts the nozzle into its mouth so that the aperture is blocked; he then inflates the vessel by blowing through the tube.

Spechter (German): A drinking glass made in the Spessart region of Germany in the 16th century. Although the term is frequently applied to a tall glass resembling a ***Stangenglas,*** there is no proof that the usage is correct.

Spill holder: A tall, narrow vessel for spills. Spills are thin

Stained glass. U.S., about 1910-1920. OH. 119.5 cm.

strips of wood, or folded or twisted pieces of paper, used for lighting candles, pipes, etc. See also **Taperstick**.

Spoon holder: A tall, narrow container for spoons. Spoon holders were used at the table from the mid-19th century to World War I.

Sport cup: A first-century A.D. Roman **mold-blown** drinking vessel decorated with fighting gladiators or a chariot race.

Sprinkler flask: A vessel with a narrow neck, sometimes with a diaphragm at the bottom, that causes the contents to emerge drop by drop.

Stained glass: The generic name for decorative windows made of pieces of colored glass fitted into **cames** and set in iron frames. Strictly speaking, the term is inaccurate because, in addition to glass colored by **staining**, glaziers used, and continue to use, glass colored throughout by **metallic oxide**, glass colored by **flashing**, and glass decorated with **enamel**.

Staining: In glassworking, the process of coloring the surface of glass by the application of silver sulfide or silver chloride, which is then fired at a relatively low temperature. The silver imparts a yellow, brownish yellow, or ruby-colored stain, which may be painted, engraved, or etched.

Stangenglas (German, "pole glass"): A tall, narrow, cylindrical drinking vessel (hence the name "pole glass"), usually with a pedestal foot.

Stemware: The collective name for drinking vessels and serving dishes with a stem supporting the bowl.

Stick-lighting, stickwork: The process of using a point to scratch internal details in painted or enameled decoration.

Stippling. England, engraved by Laurence Whistler, 1974. H. 22.4 cm.

Stippling: (1) The technique of tapping the surface of a glass object with a pointed tool, often with a diamond or tungsten-carbide tip. Each tap produces a mark, and the decoration is composed of many hundreds or thousands of marks. (2) On **lacy-pattern** pressed glass, the stippling is part of the decoration of the **mold**.

Stirrup cup: (1) A cup of wine or some other drink handed to a person when on horseback and about to set out on a journey; hence (2) a drinking vessel for consuming such drinks.

Stone: An accidental inclusion in glass. Stones consist of unmelted particles of **batch**, fragments of **refractory** material from the **pot**, or **devitrification** crystals. Stones of the first two varieties are generally irregular but rounded; those of the third variety are angular and well formed.

Strain cracks: Fissures in the body of a vessel caused by internal strain resulting from inadequate **annealing** and/or accidental thermal shock.

Strigil (Latin *strigilis*, "scraper"): A scraper used by the ancient Romans to remove impurities from the skin after bathing. Although they were usually made of metal, a few glass strigils are known to exist.

Striking: The process of reheating glass after it has cooled, in order to develop color or an opacifying agent that appears only within a limited range of temperatures.

Studio glass: A term popularized in the 1960s for unique or limited-edition objects designed and made in a studio rather than a factory, often, but not necessarily, by the same person.

Studio Glass Movement: A movement that began in the United States in the 1960s and has spread all over the world. It is characterized by the proliferation of glass artists who are not affiliated with factories, but work with hot glass in their own studios. The emergence of independent glass artists was made possible by Harvey Littleton and Dominick Labino's development in 1962 of a small **furnace** and easy-to-melt glass.

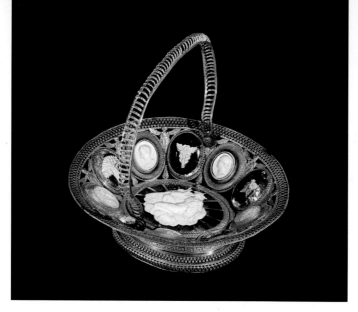

Basket with sulphides. England, about 1820-1830. OH. 20 cm.

Sturzbecher (German, "fall down beaker"): See **Cone beaker**.

Sulphide (French *sulfure*, "sulfide"): A small ornamental object of white porcelainlike material, made to be encased in glass. The term is also applied to objects that are decorated with sulphides. They were popular in Europe and America throughout the 19th century. It is not known why these objects are called sulphides; in chemistry, a sulfide is a compound containing the element sulfur.

Sweetmeat: A vessel for serving sweetmeats, sweet food such as preserved or candied fruit and sugared cakes or nuts.

Swirled ribbing: A pattern of spiraling vertical ribs made by inflating the **parison** in a **dip mold** with vertical ribs, withdrawing it, and twisting it before continuing the process of inflation. The pattern is also described as **wrythen**.

T

Table set: A group of matching objects comprising a sugar bowl, **creamer**, **spoon holder**, and butter dish.

Tank: A large receptacle constructed in a **furnace** for melting the **batch**.

Taperstick: A tall, thin vessel for tapers. Tapers are long wicks coated with wax for use as a spill. See also **Spill holder**.

Tazza (Italian, "cup"): An ornamental dish or cup on a stemmed foot. *Tazza*s were generally made for displaying fruit or sweetmeats, or as purely decorative objects.

Teardrop: A drop-shaped air bubble enclosed in a glass, usually in the stem.

Tessera (Latin, "small square tablet or block"): A small piece of glass or other suitable material, used in the formation of mosaics.

Tieback: A device for holding a drawn curtain back from the window. Some 19th- and 20th-century tiebacks have glass pommels or bosses.

Tiefschnitt (German, "deep cut"): See **Intaglio**.

Toddy plate: A popular term for a small **pressed glass** plate, made between about 1830 and 1870, presumably as a saucer under a toddy glass. Toddy is a beverage composed of whiskey or another liquor, hot water, and sugar.

Toilet bottle: A bottle for perfume or toilet water. Ancient Roman toilet bottles are frequently known as **unguentaria**.

Tool: (Noun) Any instrument used by glassworkers to develop and shape an object. Glassworkers' tools include the **blowpipe, pontil, gathering iron, pucellas, shears, clapper, pallet, block, pincers, battledore, lipper,** and **crimper.** (Verb) To alter an object with a tool.

Trail: A strand of glass, roughly circular in section, drawn out from a **gather.**

Trailing: The process of applying **trails** of glass as decoration on the body, handle, or foot of a vessel. It is done by laying or winding softened threads onto a glass object. See also **Combed decoration.**

Trick glass: A glass, usually for wine and often of extraordinary shape, designed to be as difficult as possible to drink from without spilling the contents. In drinking competitions, any drinker who spilled wine was required to start again with a full glass.

Trulla (Latin): The popular term for an ancient Roman dipper in the form of a shallow bowl with a single horizontal handle.

Tureen (French *terrine,* "flat-bottomed dish"): A deep, usually oval bowl with a lid, for serving soup; also, a smaller vessel with the same form, for serving sauce or gravy.

Twist: A type of decoration in the stems of 18th-century and later drinking glasses, made by twisting a glass **rod** embedded with threads of white or colored glass, columns of air (**air twists**), or a combination of all three.

U

Undercutting: The technique of decorating glass in high relief by cutting away part of the glass between the body of an object and its decoration (e.g., on a **cage cup**).

Glasses with twists. England, about 1760-1780.
H. (largest) 20.0 cm.

Unguentarium (Latin, pl. unguentaria): A term
commonly applied to ancient Roman **toilet bottles**. It
appears, however, that the term was "invented" in the 19th
century, on analogy with *unguentarius* ("perfume seller")
and similar Latin words that the Romans used in
connection with perfumes.

Uranium glass: Glass colored with uranium oxide. This
brilliant yellow-green glass was first made in the 1830s.

V

Vermiculée design (French, "vermiculate"): A
convoluted ground pattern resembling worm tracks.

Verre églomisé (French): A decorative technique in
which gold or silver leaf is applied to the back side of a
piece of glass, engraved, and protected by varnish, metal
foil, or another piece of glass. The name is derived from

the French mirror and picture framer Jean-Baptiste Glomy (d. 1786). Decoration of this type, however, had been made since the 13th century and the term **reverse foil engraving** is preferable.

Vetro a fili (Italian, "glass with threads"): A type of blown glass made with **canes** that form a pattern of parallel lines.

Vetro a reticello (Italian, "glass with a small network"): A type of blown glass made with canes laid in a crisscross pattern to form a fine net, which may contain tiny **air traps**.

Vetro a retorti (Italian, "glass with twists"): A type of blown glass made with **canes** that have been twisted to form spiral patterns.

Vetro di trina (Italian, "lace glass"): A term loosely applied to various types of ***vetro a reticello***.

Victory beaker: A first-century A.D. Roman mold-blown drinking vessel inscribed in Greek with words meaning "Take the victory."

W

Waldglas (German, "forest glass"): See **Forest glass**.

Warzenbecher (German, "wart beaker"): A heavy glass tumbler made of **forest glass** and decorated with prunts. It was produced in Germany in the 16th and 17th centuries.

Waster: A defective object discarded during manufacture. Wasters are routinely recycled as **cullet**.

Water set: A pitcher with matching tumblers, sometimes with a matching tray.

Wheel engraving. U.S., 1935. OH. 17.7 cm.

Weathering: Changes on the surface of glass caused by chemical reaction with the environment. Weathering usually involves the leaching of **alkali** from the glass by water, leaving behind siliceous weathering products that are often laminar.

Wheel engraving: A process of decorating the surface of glass by the grinding action of a wheel, using disks of various materials (usually copper, but sometimes stone) and sizes, and an abrasive in a grease or slurry applied to a wheel, as the engraver holds the object against the underside of the rotating wheel. See **Copper-wheel engraving**.

Whimsy: An object made solely for fun or to show off the skill of the glassworker.

Whiskey set: A decanter with matching tumblers, sometimes with a matching tray.

Williamite glass: A 17th-century English drinking vessel engraved with a toast, a symbol (an orange tree, for example), or a motto supporting King William III, or his portrait. William III, who was the hereditary prince of Orange, came to the English throne in 1689. His political opponents were the Jacobites; see **Jacobite glass**.

Wine set: A decanter with matching wineglasses, sometimes with a matching tray.

Winged goblet: A type of **goblet** with the stem in the form of vertical, winglike flanges composed of trails arranged in a complex design that may include dragons, sea horses, and other creatures. The German term for a winged goblet is *Flügelglas*.

Witch ball: A glass globe intended to be hung in a prominent place to ward off the evil eye.

Wrythen: See **Swirled ribbing**.

Y

Yard-of-ale: A type of ale glass with a trumpet-shaped mouth, a long and narrow neck, and a small, globular body. So-called because they were often one yard (three feet) long, these objects contained about one pint and functioned like **trick glasses**. The yard-of-ale is an English form, which came into use in the 1680s and continued into the 19th century.

Zwischengoldglas *beaker. Bohemia, about 1748. H. 9.5 cm.*

Z

Zwischengoldglas (German, "gold between glass"): A type
of decoration, produced in Bohemia and Austria in the
18th century, in which a design in gold or silver leaf is
incorporated between two vessels that fit together
precisely. Unlike Hellenistic and Roman **gold glass**,
which is fused, ***Zwischengoldglas*** is bonded with cement.

DESCRIPTIONS OF ILLUSTRATED OBJECTS

(The number preceding each description below refers to the page on which the object is illustrated.)

4. Acid-etched vase. France, Bar-sur-Seine, made by Maurice Marinot, 1934. H. 17.1 cm. Gift of Mlle Florence Marinot.

7. Amphoriskos. Eastern Mediterranean, 2nd-1st century B.C. H. 24 cm.

9. Gold Aurene vase. U.S., Corning, NY, Steuben, about 1910. H. 17.1 cm. Gift of Corning Glass Works.

10. Baluster. England, about 1695. H. 24.7 cm.

12. Beakers. Austria, Vienna, enameled by Anton Kothgasser and Gottlob Samuel Mohn, about 1811-1828. H. (tallest) 11 cm. Gift of Mrs. K. F. Landegger.

15. Burmese lamp. U.S., Mount Washington Glass Company, about 1885-1895. H. 48.5 cm. Part gift of William E. Hammond.

17. Cameo glass plaque. England, Thomas Webb & Sons, carved by George Woodall, 1898. D. 46.3 cm. Bequest of Mrs. Leonard S. Rakow.

19. Chair. From Denis Diderot, *Receuil de planches sur les sciences...*, edition of about 1775.

21. *Der Tempel von Adam und Eva*. U.S., Robert Carlson, 1986. H. 64.5 cm. Gift of Raymond E. Fontaine.

22. Mug. U.S., probably New Jersey, about 1860-1870. H. 15.3 cm. Gift of The Ruth Bryan Strauss Memorial Foundation.

25. Making crown glass. From Denis Diderot, *Receuil de planches sur les sciences...*, edition of about 1775.

27. *Daumenglas*. Germany, 17th century. OH. 36.8 cm.

28. Dish. France or the Netherlands, mid-17th century. D. 48.8 cm. Museum Endowment Fund purchase.

30. Sprinkler bottle and tumblers. Venice, probably Miotti, about 1730-1735. H. (bottle) 27.5 cm.

33. Flameworked figure of a beggar. France, probably Nevers, early 18th century. H. 5.0 cm.

35. Forest glass *Nuppenbecher*. Germany, 15th-16th century. H. 7.8 cm.

37. Toilet water bottle. England, probably Bristol, about 1790. H. 13.1 cm.

38. Gold-glass roundel. Roman Empire, Italy, 4th century A.D. D. 9.7 cm.

40. Head flask. Roman Empire, eastern Mediterranean, 4th century A.D. H. 19.6 cm.

43. Jack-in-the-pulpit vase. U.S., Tiffany Studios, 1912. H. 47.6 cm.

45. Kohl tube. Achaemenian Empire, Iran, 5th-4th century B.C. H. 8.8 cm.

46. Lacy-pattern tray. U.S., New England, about 1830-1845. L. 23.8 cm.

48. Covered sugar bowl and creamer. U.S., New York State, about 1835-1850. OH. (bowl) 27.2 cm.

51. Using the marver. From Denis Diderot, *Receuil de planches sur les sciences...*, edition of about 1775.

53. Glassworker's mold, copper alloy. Islamic, western Asia, 11th-13th century. H. 11.4 cm.

55. *Frosted Radio Light* from the "Spiral Series." U.S., Paul Seide, 1986. H. 48.4 cm. Part gift of Mike Belkin.

56. Goblet. England, London, Savoy glasshouse of George Ravenscroft, 1676-1678. H. 18.6 cm.

59. Paperweight. France, Paris, Cristallerie de Pantin, 1870-1880. D. 11.5 cm. Gift of The Hon. and Mrs. Amory Houghton.

60. "Morgan Vase." U.S., Wheeling, WV, about 1886-1891. OH. 25.5 cm.

63. *Porróns*. Spain, Catalonia, 18th century. H. (larger) 27.2 cm. Bequest of Jerome Strauss and gift of The Ruth Bryan Strauss Memorial Foundation.

65. Punch set. France, Cristalleries de Baccarat, 1867. H. 56.5 cm. Gift of Mrs. Charles K. Davis.

67. *Reichsadler Humpen*. Bohemia, 1574. H. 26.4 cm. Gift of Edwin J. Beinecke.

68. Ribbon glass cup. Roman Empire, probably Italy, 1st century B.C.-1st century A.D. H. 4.8 cm.

71. *Schwarzlot* tumbler. Germany, early 18th century. H. 9.5 cm.

72. Part of service. Austria, Wiener Werkstätte, designed by Josef Hoffmann, about 1916. H. (tallest) 32.8 cm.

75. *Dante and Beatrice*. U.S., Philadelphia, PA, Willet Studios, about 1910-1920. OH. 119.5 cm. Gift of Dr. Thomas H. English.

76. *The Overflowing Landscape*. England, engraved by Laurence Whistler, 1974. H. 22.4 cm.

78. Basket. England, London, Falcon Glassworks of Apsley Pellatt, about 1820-1830. OH. 20 cm. Gift of Mr. and Mrs. Paul Jokelson.

81. Drinking glasses. England, about 1760-1780. H. (largest) 20.0 cm. Group includes bequest of Jerome Strauss and gifts of The Ruth Bryan Strauss Memorial Foundation.

83. *Gazelle Bowl*. U.S., Corning, NY, Steuben Glass, Inc., designed by Sidney Waugh, 1935. OH. 17.7 cm. Gift of Mr. and Mrs. John K. Olsen.

85. Beaker with the arms of an abbot of Vyssi Brod. Bohemia, about 1748. H. 9.5 cm. Bequest of Jerome Strauss.